Constructing a Segment Congruent to a Given Segment

A ——————————— B

1. Draw a circle with radius $\overline{AB}$.

2. Draw a circle with center C and radius congruent to $\overline{AB}$.

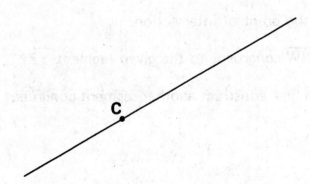

3. Label as D a point of intersection of this circle and the line.

4. Is $\overline{CD}$ congruent to $\overline{AB}$? _ _ _ _ _

4

Problem: *On a line, <u>construct</u> a segment congruent to a given segment.*

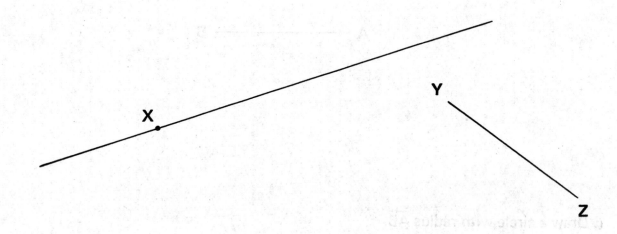

Solution:

1. Draw an arc with center X and radius congruent to $\overline{YZ}$ which intersects the given line.

2. Label as W the point of intersection.

   Is segment $\overline{XW}$ congruent to the given segment $\overline{YZ}$? _ _ _ _ _

3. On the given line, construct another segment congruent to $\overline{YZ}$.

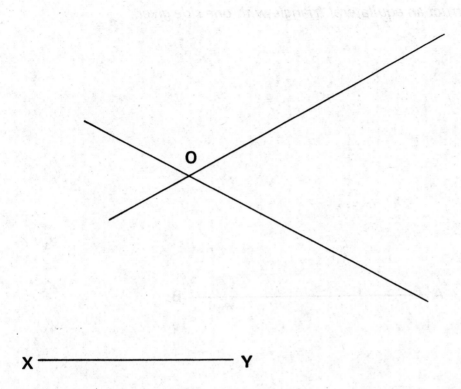

1. On each given line, construct a segment congruent to $\overline{XY}$ and with O as one endpoint.

2. Label as A and B the other endpoints of these two segments.

3. Draw the segment determined by A and B.

4. Is the triangle AOB equilateral? _ _ _ _ _

Constructing Equilateral Triangles

Problem: *Construct an equilateral triangle with one side given.*

A ——————————————————————————— B

Solution:

1. Draw an arc above $\overline{AB}$ with center A and radius $\overline{AB}$.

2. Draw an arc above $\overline{AB}$ with center B and radius $\overline{AB}$.
   Make the two arcs intersect.

3. Label as C the point of intersection.

4. Draw triangle ABC.

5. What kind of triangle is ABC? _ _ _ _ _ _ _ _ _ _ _ _ _ _

   How do you know? _

Key to
Geometry®

CONSTRUCTIONS

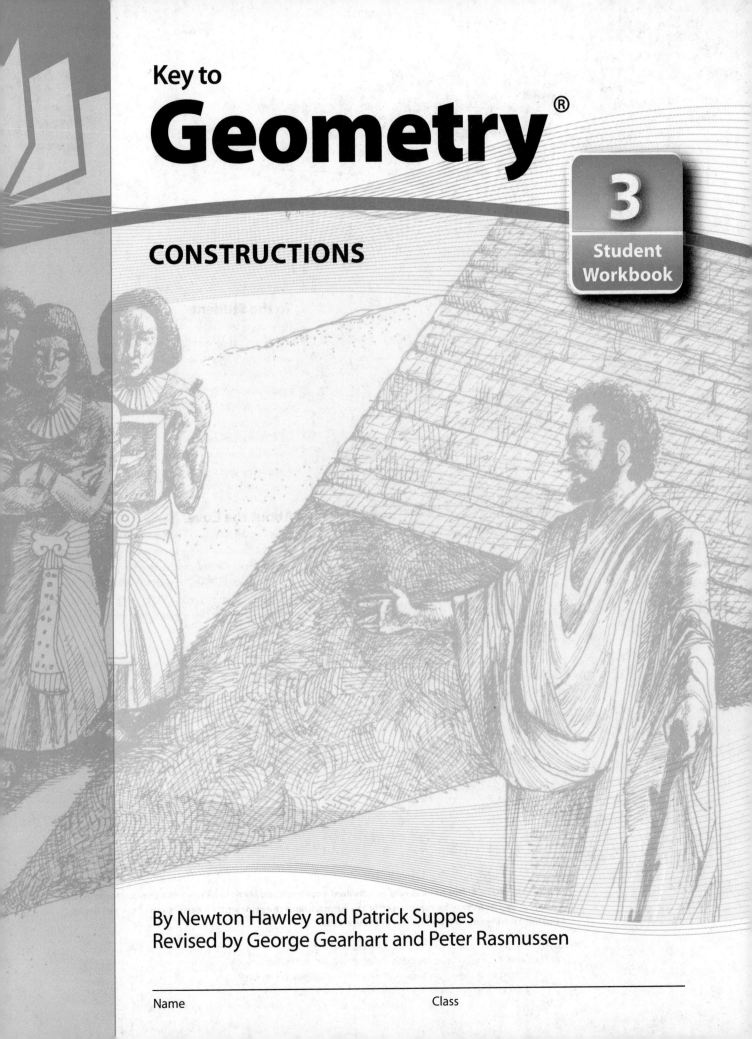

3
Student
Workbook

By Newton Hawley and Patrick Suppes
Revised by George Gearhart and Peter Rasmussen

Name _____ Class _____

Key to
Geometry® CONSTRUCTIONS

3
Student
Workbook

TABLE OF CONTENTS

To the Student

These books will help you to discover for yourself many important relationships of geometry. Your tools will be the same as those used by the Greek mathematicians more than 2000 years ago. These tools are a *compass* and a *straightedge*. In addition, you will need a *sharpened pencil*. The lessons that follow will help you make drawings from which you may learn the most.

The answer books show *one* way the pages may be completed correctly. It is possible that your work is correct even though it is different. If your answer differs, re-read the instructions to make sure you followed them step by step. If you did, you are probably correct.

About the Cover

Thales of Miletus was a Greek philosopher. He lived around 600 B.C., over 2500 years ago. Before Thales people thought of geometry only in concrete and practical ways. A line was a stretched rope or the edge of a field. Thales began the Greek tradition of studying the principles of geometry separate from the application of geometry in the physical world. Thales traveled widely in Egypt and the Middle East where he studied Egyptian mathematics. When he returned to Greece he brought many Egyptian discoveries in geometry back with him.

On the cover of this booklet Thales uses his knowledge of geometry to measure the height of a pyramid. By placing a stick in the ground and measuring the height of the stick, the length of the stick's shadow and the length of the pyramid's shadow, Thales can calculate the height of the pyramid.

Cover art by Howard Coale

Copyright © 1979, 1972 by Holden Day, Inc. All rights reserved.
® *Key to Fractions, Key to Decimals, Key to Percents, Key to Algebra, Key to Geometry, Key to Measurement,*
and *Key to Metric Measurement* are registered trademarks of Key Curriculum Press.
Published by Key Curriculum Press, 1150 65th Street, Emeryville, CA 94608
ISBN 978-0-913684-73-3   10 LOV 22   Printed in the United States of America

1. Draw an arc <u>above</u> $\overline{RS}$ with center R and radius $\overline{RS}$.

R ———————————————————— S

Then draw an arc above $\overline{RS}$ with center S and radius $\overline{RS}$.

Make the arcs intersect.

Label as T the point of intersection.

2. Draw the segment determined by R and T.

3. Draw $\overline{ST}$.

4. What kind of triangle is RST? _ _ _ _ _ _ _ _ _ _ _ _ _

5. Name the angles of triangle RST.

_ _

1. Draw $\overline{AB}$.

   Then draw intersecting arcs with center A and radius $\overline{AB}$, and with center B and radius $\overline{AB}$.

B .

A .

2. Construct an equilateral triangle with $\overline{AB}$ as one side.

3. Construct another equilateral triangle below.

9

1. Construct an equilateral triangle with side $\overline{AB}$.

A ————————————————— B

2. Label as C the vertex of this triangle.

3. Now construct an equilateral triangle with side $\overline{AC}$ and vertex D outside triangle ABC.

4. Construct an equilateral triangle with side $\overline{BC}$ and vertex E outside triangle ABC.

5. Draw $\overline{DB}$.

6. Is triangle ADB equilateral? _ _ _ _ _

7. Draw $\overline{DE}$.

8. Is triangle DBE equilateral? _ _ _ _ _

1. On the given line, construct a segment congruent to $\overline{AB}$.

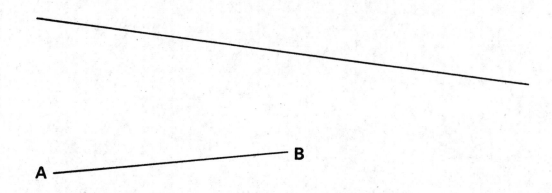

2. Construct an equilateral triangle with the segment $\overline{SY}$ as one side.

3. On the given line, construct a segment congruent to $\overline{GH}$ and with K as one endpoint.

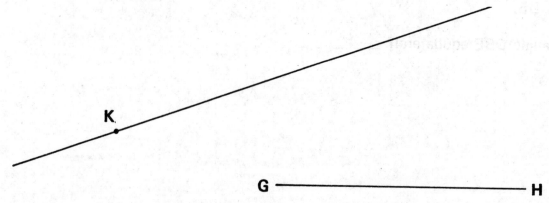

4. Label as L the other endpoint of the new segment.

Review

1. Segment $\overline{AB}$ is _ _ _ _ _ _ _ _ _ _ _ _ _ _ _ _ _ $\overline{AC}$.

   (a) shorter than

   (b) congruent to

   (c) longer than

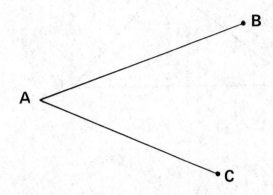

2. Draw a line.

   On the line, construct a segment congruent to $\overline{XY}$.

3. Construct an equilateral triangle.

Constructing a Triangle Congruent to a Given Triangle

Problem: *To construct a triangle congruent to a given triangle.*

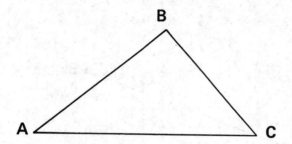

_____

Solution:

1. On the line, construct a segment congruent to $\overline{AC}$.
   Label the endpoints P and Q.

2. Draw an arc with center P and radius congruent to $\overline{AB}$.

3. Draw an arc with center Q and radius congruent to $\overline{BC}$.
   Make the two arcs intersect.

4. Label as R the point where the two arcs intersect.

5. Draw triangle PQR.

6. Which side is congruent to $\overline{AB}$? _ _ _ _ _

7. Which side is congruent to $\overline{AC}$? _ _ _ _ _

8. Which side is congruent to $\overline{BC}$? _ _ _ _ _

9. Are the triangles congruent? _ _ _ _ _

1. Draw a line and on it construct a segment congruent to $\overline{EF}$. Label its endpoints X and Y.

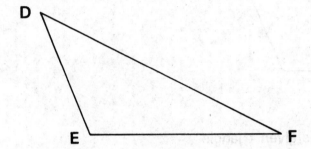

2. Draw an arc with center X and radius congruent to $\overline{DE}$.

3. Draw an arc with center Y and radius congruent to $\overline{DF}$. Make the two arcs intersect.

4. Label the point of intersection Z.

5. Draw triangle XYZ.

6. Is triangle XYZ congruent to triangle DEF? _ _ _ _ _

7. Construct another triangle congruent to triangle DEF.

1. Are all four sides of the quadrilateral congruent? _ _ _ _ _

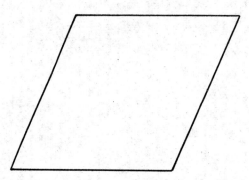

2. Construct a triangle congruent to the given triangle.

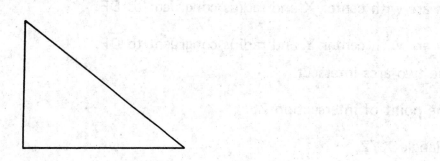

3. Construct a triangle congruent to the given triangle.

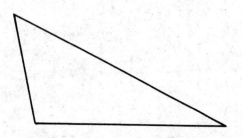

1. Compare segment $\overline{AB}$ with segment $\overline{DE}$.

   Are they congruent? _ _ _ _ _

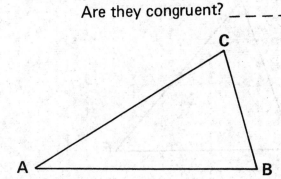
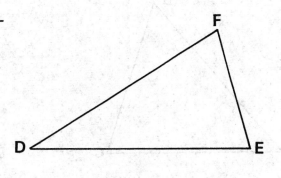

2. Which side is congruent to $\overline{AC}$? _ _ _ _ _
   Compare them.

3. Which side is congruent to $\overline{BC}$? _ _ _ _ _
   Compare them.

4. Is triangle ABC congruent to triangle DEF? _ _ _ _ _

5. Compare the sides of triangle RST to the sides of triangle WXY.

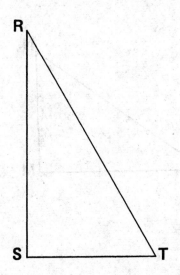
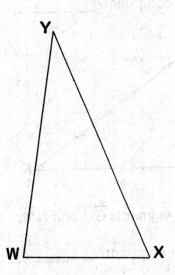

6. Are the triangles congruent? _ _ _ _ _

7. How do you know? _

   _

1. Trace triangle ABC.

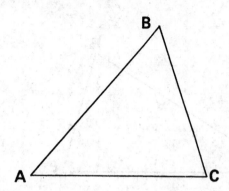 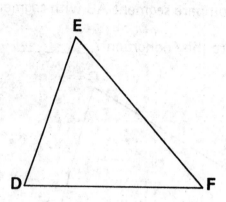

2. Place your tracing over triangle DEF.

   Can you make the triangles match exactly? _ _ _ _ _

3. Turn your tracing over.

   Now can you make the triangles match? _ _ _ _ _

4. Are triangles ABC and DEF congruent? _ _ _ _ _

5. Compare segment $\overline{HK}$ with segment $\overline{LM}$.

   Are they congruent? _ _ _ _ _

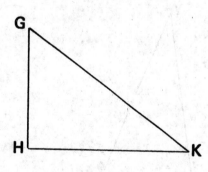 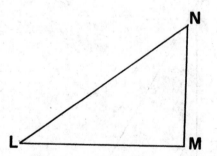

6. Compare segments $\overline{GH}$ and $\overline{NM}$.

   Are they congruent? _ _ _ _ _

7. Compare segments $\overline{GK}$ and $\overline{LN}$.

   Are they congruent? _ _ _ _ _

8. Is triangle GHK congruent to triangle LMN? _ _ _ _ _

Doubling Segments

X .   . Y

1. Draw a long line through X and Y.

2. Draw a circle with center Y passing through X.

3. Label as Z the other point of intersection of the circle and the line.

4. Is segment $\overline{XY}$ congruent to the segment $\overline{YZ}$? _ _ _ _ _

5. Is segment $\overline{XZ}$ <u>double</u> the segment $\overline{XY}$? _ _ _ _ _

Problem: *Double a given line segment.*

A •————————————————• B

Solution:

1. Draw a long line through A and B.

2. Draw an arc with center B and radius $\overline{AB}$ which intersects the long line to the right side of B (but not at A).

3. Label as C the point of intersection.

   Is segment $\overline{AC}$ double the length of $\overline{AB}$? _ _ _ _ _

4. Double the line segment below.

1. Construct an equilateral triangle with sides double the segment $\overline{AB}$.

A •————————————• B

2. Can you construct an equilateral triangle with side $\overline{XY}$ and vertex

at point Z? _ _ _ _ _

Z
•

X ————————————————————— Y

1. Double the given segments.

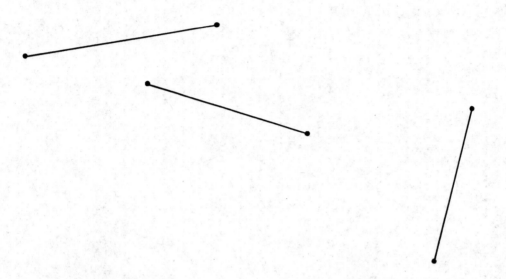

2. Construct an equilateral triangle with side $\overline{AB}$.

A

B

3. On the given line, construct a segment congruent to $\overline{UV}$.

U ——————————————— V

Tripling Segments

1. Draw a long line through A and B.

B

A

2. Draw a circle with center B and radius $\overline{AB}$.

3. Label as C the <u>other</u> intersection of this circle and the line.

4. Draw a circle with center C and radius $\overline{BC}$.

5. Label as D the other intersection of this circle and the line.

6. How many <u>times as long as</u> $\overline{AB}$ is the segment $\overline{AD}$? _ _ _ _ _ _ _ _

    (a) two times as long.   (c) four times as long.

    (b) three times as long.   (d) five times as long.

Problem: *Triple a given segment.*

A •————————————• B

Solution:

1. Double segment $\overline{AB}$.

2. Label as C the <u>right</u> endpoint of this new segment.

3. Draw an arc with center C and radius $\overline{BC}$ which intersects the line on which points A, B, and C lie.
   Draw this arc to the right of C.

4. Label as D this new point of intersection.

   Is segment $\overline{AD}$ three times as long as or triple the length of

   $\overline{AB}$? _ _ _ _ _

5. Triple the segment given below.

R •————————————• S

1. Segment $\overline{AB}$ is _ _ _ _ _ _ _ _ _ _ _ _ _ _ _ _ _ double the length of segment $\overline{XY}$.

   (a) <u>less than</u>

   (b) <u>exactly</u>

   (c) <u>more than</u>

2. Segment $\overline{AB}$ is _ _ _ _ _ _ _ _ _ _ _ _ _ _ _ _ triple the length of segment $\overline{XY}$.

   (a) less than

   (b) exactly

   (c) more than

3. Construct a triangle congruent to triangle LMN.

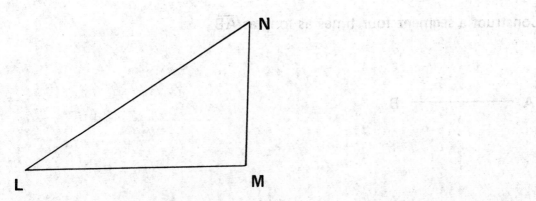

1. Double the given segments. Make each segment twice as long.

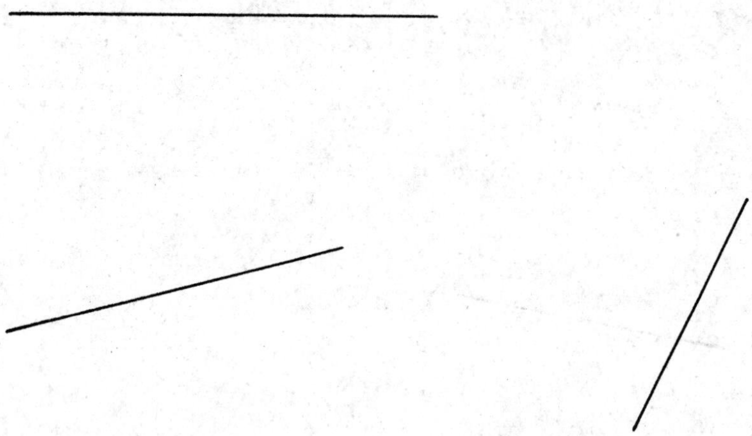

2. Construct segments three times as long as or triple the given segments $\overline{AZ}$ and $\overline{PU}$.

A ——————— Z

P ——————— U

3. Construct a segment four times as long as $\overline{AB}$.

A ——————— B

1. On the given line, construct a segment congruent to $\overline{AB}$.

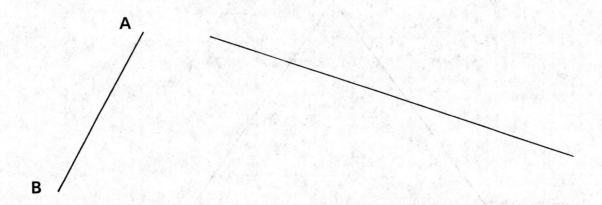

2. Double segment $\overline{CD}$.

3. Triple segment $\overline{EF}$.

4. Construct a segment five times as long as $\overline{GH}$.

Review

1. Is triangle XYZ an equilateral triangle? _ _ _ _ _

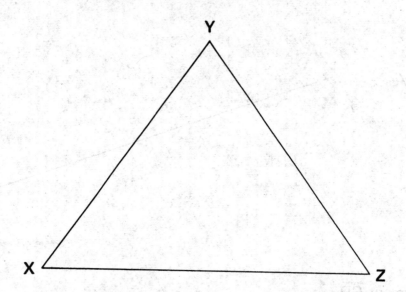

2. Construct a triangle congruent to the given triangle.

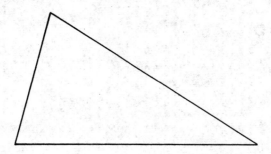

3. Construct an equilateral triangle with $\overline{AB}$ as one side.

4. Construct another equilateral triangle with $\overline{AB}$ as one side.

Bisecting Segments

Problem: _Bisect a given segment._

A ——————————— B

Solution:

1. Draw a circle with center A and radius $\overline{AB}$.

2. Draw a circle with center B and radius $\overline{AB}$.

3. Label as C and D the points where the circles intersect.

4. Draw $\overline{CD}$.

5. Label as E the point of intersection of $\overline{AB}$ and $\overline{CD}$.

6. Now fold this page along $\overline{CD}$.

   Does $\overline{AE}$ fit exactly over $\overline{BE}$? _ _ _ _ _

   Is $\overline{AE}$ congruent to $\overline{BE}$? _ _ _ _ _

   Segment $\overline{CD}$ bisects segment $\overline{AB}$.

   Point E is the _midpoint_ of $\overline{AB}$.

A ——————————— B

1. Draw a circle with center A and radius $\overline{AB}$.

2. Draw a circle with center B and radius $\overline{AB}$.

3. Label as C and D the points of intersection of these circles.

4. Draw $\overline{CD}$.

5. Label as E the point of intersection of $\overline{AB}$ and $\overline{CD}$.

6. Compare $\overline{AE}$ and $\overline{EB}$.

7. Does point E bisect $\overline{AB}$? _ _ _ _ _

8. Point E is the _ _ _ _ _ _ _ _ _ _ _ _ _ _ _ _ of segment $\overline{AB}$.
   (a) bisect
   (b) midpoint
   (c) half
   (d) congruent

A ——————————————— B

1. Draw arcs above and <u>below</u> $\overline{AB}$ with center A and radius $\overline{AB}$.

2. Draw arcs with center B and radius $\overline{AB}$ above and below $\overline{AB}$. Make the arcs intersect.

3. Label as C and D the points of intersection of the arcs.

4. Draw segment $\overline{CD}$.

5. Label as E the point of intersection of $\overline{AB}$ and $\overline{CD}$.

6. Is $\overline{AE}$ congruent to $\overline{BE}$? _ _ _ _ _ _

7. $\overline{CD}$ _ _ _ _ _ _ _ _ _ _ _ _ _ _ $\overline{AB}$.

   (a) bisects   (c) halves

   (b) midpoint   (d) makes congruent

Find the midpoint of the given segment.

A ———————————————— B

1. Draw arcs above and below $\overline{AB}$ with center A and radius $\overline{AB}$.

   Draw arcs above and below $\overline{AB}$ with center B and radius $\overline{AB}$.

   Make the arcs intersect, and label the points of intersection C and D.

2. Draw arcs above and below $\overline{AB}$ with center A and a smaller radius.

   Draw arcs above and below $\overline{AB}$ with center B and the same radius.

   Make the arcs intersect, and label the points of intersection E and F.

3. Draw $\overline{CD}$.

4. What do you notice about the points C, D, E, and F? _ _ _ _ _ _ _ _

   _

P ——————————————————— Q

1. Draw arcs above and below $\overline{PQ}$ with center P and radius <u>greater than</u> $\overline{PQ}$.

   Draw arcs above and below $\overline{PQ}$ with center Q and the same radius.

   Make the arcs intersect, and label the points of intersection R and S.

2. Draw $\overline{RS}$.

3. Label as T the intersection of $\overline{RS}$ and $\overline{PQ}$.

4. Compare $\overline{PT}$ and $\overline{TQ}$.

5. Is T the midpoint of $\overline{PQ}$? _ _ _ _ _

6. Compare $\overline{RT}$ and $\overline{TS}$.

7. Is T the midpoint of $\overline{RS}$? _ _ _ _ _

C ——————————————————— D

1. Draw arcs above and below $\overline{CD}$ with center C and radius <u>smaller</u> <u>than</u> $\overline{CD}$.

   Draw arcs above and below $\overline{CD}$ with center D and the same radius.

   Make the arcs intersect, and label the points of intersection E and F.

2. Draw $\overline{EF}$.

3. Label as G the intersection of $\overline{CD}$ and $\overline{EF}$.

4. Compare $\overline{CG}$ and $\overline{GD}$.

5. Is G the midpoint of $\overline{CD}$? _ _ _ _ _

6. Does $\overline{EF}$ bisect $\overline{CD}$? _ _ _ _ _

1. Bisect the given segments.

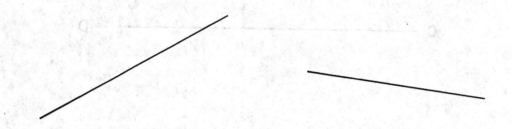

2. Double segment $\overline{AB}$.

A ——————————————— B

3. On segment $\overline{UV}$ construct a segment congruent to $\overline{XY}$.

U ——————————————————————————————— V

X ——————————————— Y

1. Find the midpoint or <u>bisector</u> of segment $\overline{AB}$.

A  B

2. Are segments $\overline{CD}$ and $\overline{FG}$ congruent? _ _ _ _ _

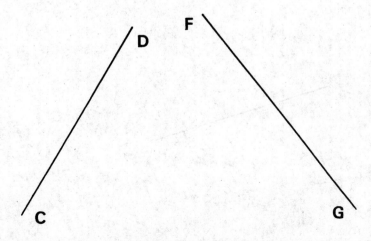

1. <u>Divide</u> segment $\overline{AB}$ into two congruent parts.

A ——————————————— B

Each part is _ _ _ _ _ _ _ _ _ _ _ _ _ _ _ _ $\overline{AB}$.

(a) twice   (c) congruent to

(b) half   (d) double

2. Divide the given segment into four congruent segments.

1. Problem: *Construct a segment 1½ times as long as $\overline{AB}$.*

A ——————————————— B

Solution:

Bisect $\overline{AB}$.

Label as E the midpoint.

Draw a long line through $\overline{AB}$.

Construct a segment congruent to $\overline{AE}$ to the right of B.

Label its endpoint F.

Is $\overline{AF}$ 1½ times as long as $\overline{AB}$? _ _ _ _ _

2. Construct a segment 1½ times as long as $\overline{CD}$.

C ——————————————— D

3. Construct a segment 2½ times as long as $\overline{CD}$.

Review

1. On the given line, construct a segment congruent to $\overline{AB}$.

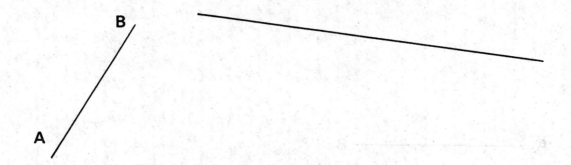

2. Construct a triangle congruent to the given triangle.

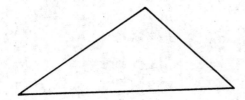

3. Is the triangle below equilateral? _ _ _ _ _

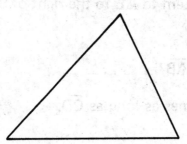

4. Bisect or divide line segment $\overline{XY}$ into two congruent parts.

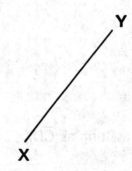

Comparing Angles

1. Extend the sides of each angle.

2. Trace angle DEF and place the tracing over angle KLM.

   Is angle DEF congruent to angle KLM? _ _ _ _ _

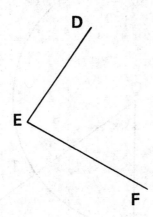

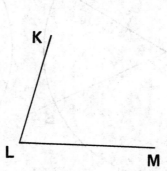

3. Are the triangles congruent? _ _ _ _ _

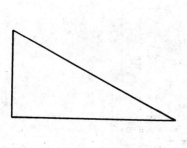

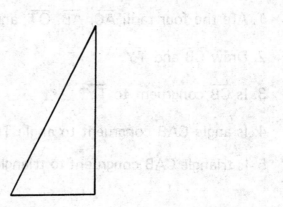

4. Are the angles congruent? _ _ _ _ _

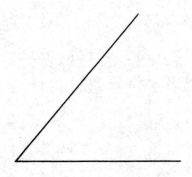

Problem: *Is angle CAB congruent to angle TOY?*

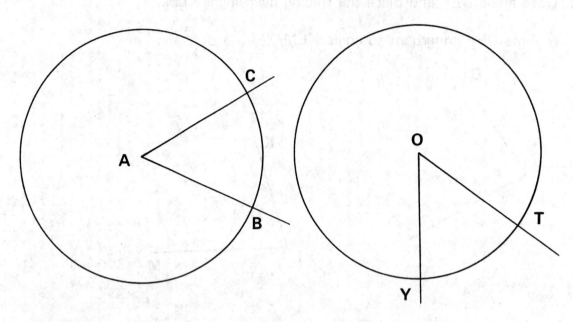

Solution:

1. Are the four radii $\overline{AC}$, $\overline{AB}$, $\overline{OT}$, and $\overline{OY}$ all congruent? _ _ _ _ _

2. Draw $\overline{CB}$ and $\overline{TY}$.

3. Is $\overline{CB}$ congruent to $\overline{TY}$? _ _ _ _ _

4. Is angle CAB congruent to angle TOY? _ _ _ _ _

5. Is triangle CAB congruent to triangle TOY? _ _ _ _ _

Problem: _Show_ that two angles are congruent.

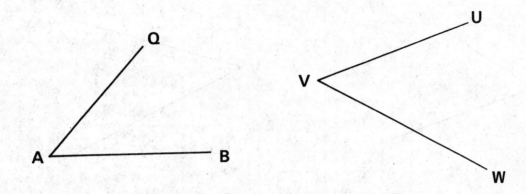

Solution:

1. Draw an arc with center A which intersects $\overrightarrow{AQ}$ and $\overrightarrow{AB}$.

2. Label as C and D the points of intersection.

3. Use the same radius to draw an arc with center V which intersects $\overrightarrow{VW}$ and $\overrightarrow{VU}$.

4. Label as X and Y the points of intersection.

5. Draw $\overline{CD}$; then draw $\overline{XY}$.

6. Is $\overline{CD}$ congruent to $\overline{XY}$? _ _ _ _ _

7. Is angle BAQ congruent to angle UVW? _ _ _ _ _

1. Draw arcs with congruent radii <u>centered at</u> A and V.

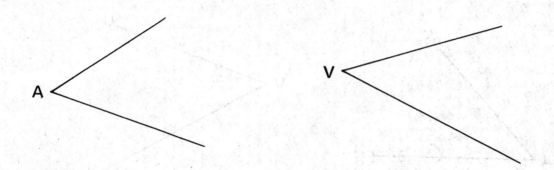

2. Label as B and C the points of intersection of the arc centered at A and the angle.

3. Label as X and Y the points of intersection of the arc centered at V and the angle.

4. Is $\overline{BC}$ congruent to $\overline{XY}$? _ _ _ _ _

5. Is angle BAC congruent to angle XVY? _ _ _ _ _

6. Are the two angles below congruent? _ _ _ _ _

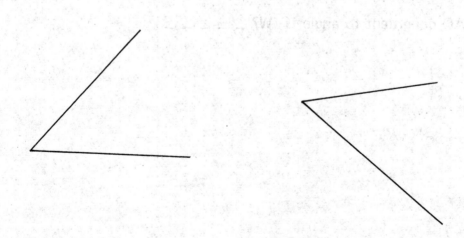

1. Show that the two angles below are congruent.

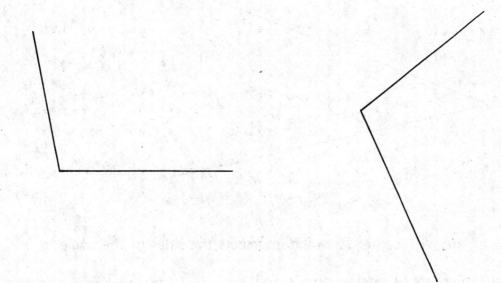

2. Bisect the segment.

3. Which angle is larger? _ _ _ _ _

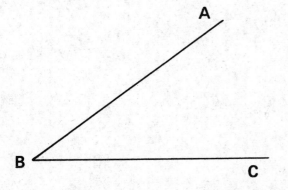

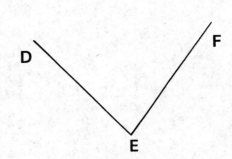

44

Bisecting Angles

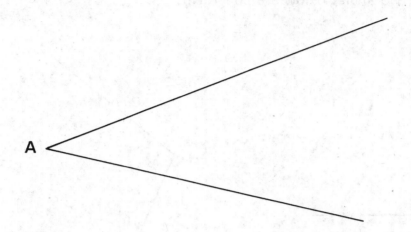

Cut along this line.

1. Draw an arc with center A which intersects the sides of the angle.

STOP.

2. Label the points of intersection X and Y.

3. Draw an arc with center X.

4. Use the same radius to draw an arc with center Y.
   Make the arcs intersect.

5. Label as Z the point of intersection of these arcs.

6. Draw the line determined by A and Z.

7. Fold your paper on $\overrightarrow{AZ}$.

   Is angle ZAX congruent to angle ZAY? _ _ _ _ _
   $\overrightarrow{AZ}$ <u>bisects</u> angle XAY.

Problem: *Bisect a given angle.*

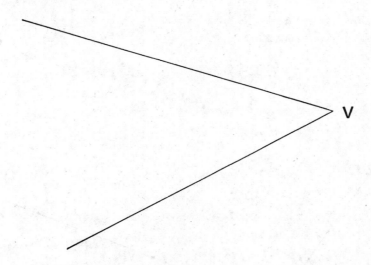

Solution:

1. Draw an arc with center V which intersects the sides of the angle.

2. Label as A and B the points of intersection.

3. Draw an arc with center A inside the angle.

4. Using the same radius, draw an arc with center B.
   Make these two arcs intersect.

5. Label as D the point of intersection of the arcs.

6. Draw the line determined by V and D.

7. Does $\overrightarrow{VD}$ bisect angle AVB? _ _ _ _ _

1. Divide angle ABC into two congruent parts.

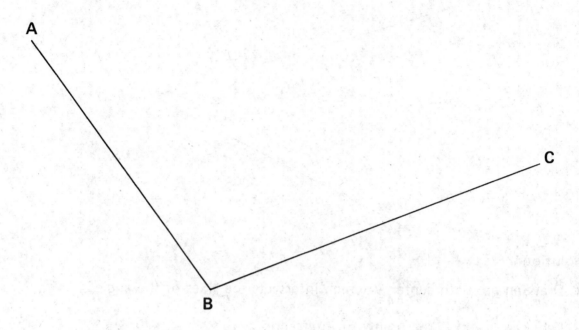

2. Bisect angle XYZ.

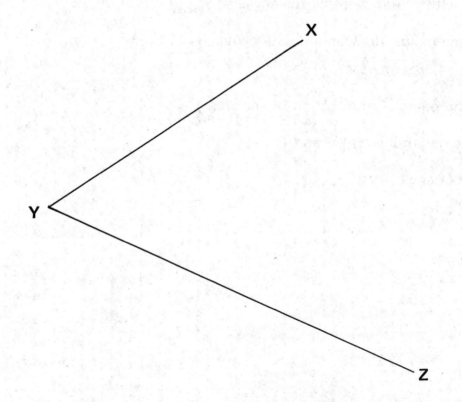

1. Bisect angle LMN.

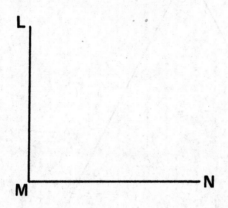

2. Bisect angle PQR.

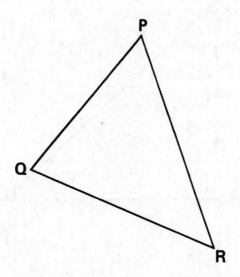

1. Bisect angle ABC.

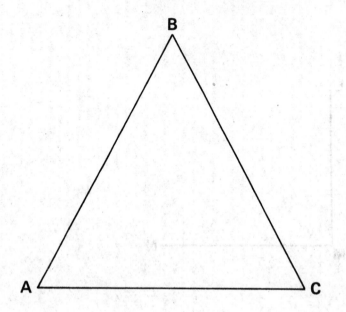

2. Bisect angle DEF.

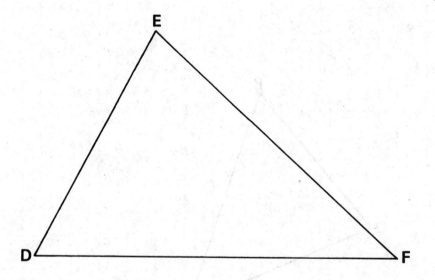

3. Bisect angle EFD.

4. Bisect angle EDF.

1. Construct an equilateral triangle with the given segment as one side.

2. Then find the midpoint of each side of the triangle.

3. Connect these midpoints.

4. How many triangles do you have? _ _ _ _ _

5. Are they all equilateral? _ _ _ _ _

Review

1. Construct an equilateral triangle with segment $\overline{AB}$ as one side.

2. On the line, construct a segment congruent to $\overline{XY}$.

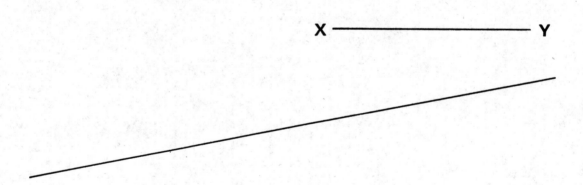

3. Bisect line segment $\overline{PQ}$.

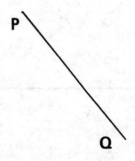

1. Show that the two angles are congruent.

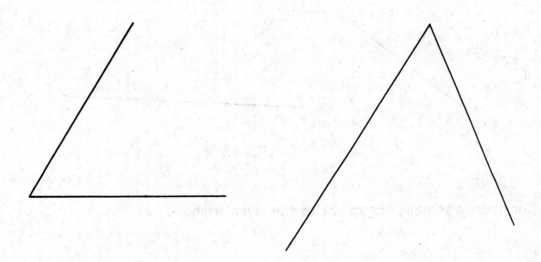

2. Bisect the angles below.

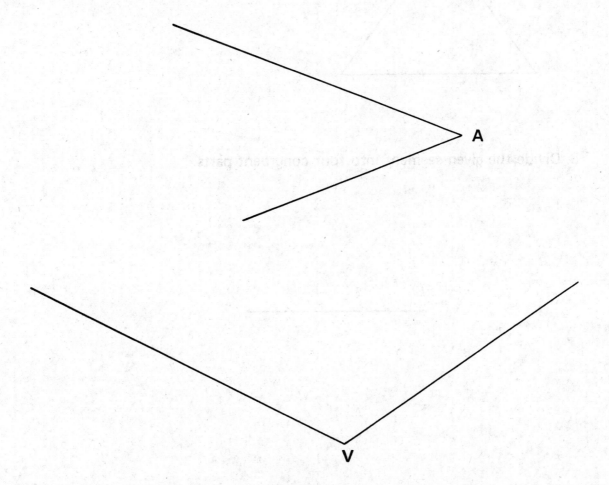

1. Are the segments congruent? _ _ _ _ _

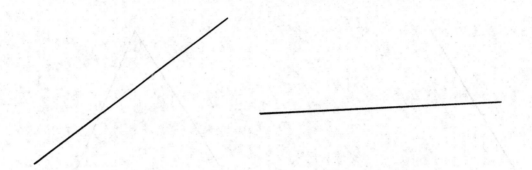

2. Construct a triangle congruent to the given triangle.

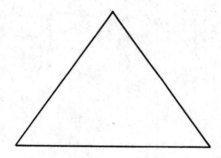

3. Divide the given segment into four congruent parts.

Practice Test

1. Construct an equilateral triangle with the given side $\overline{GH}$.

2. On the given line, construct a segment congruent to $\overline{XY}$.

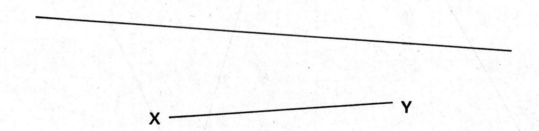

3. Double the given segment $\overline{AB}$.

A ———————————— B

4. Triple the given segment $\overline{CJ}$.

C ——————————— J

5. Bisect the given segment.

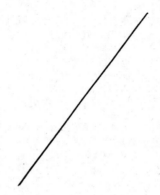

6. Are the given angles congruent? _ _ _ _ _

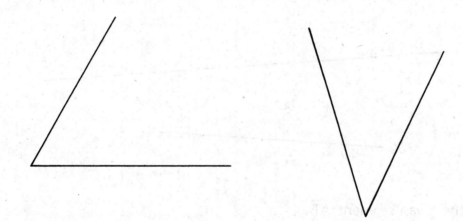

7. Bisect the given angle.

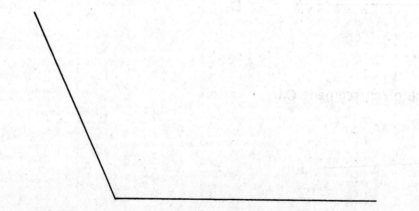

8. Two points _ _ _ _ _ _ _ _ _ _ _ _ _ _ a line.
   (a) square   (c) show
   (b) determine   (d) draw

9. The figure ABC is a _ _ _ _ _ _ _ _ _ _ _ _ _ _ .
   (a) segment   (c) triangle
   (b) angle   (d) pentagon

10. When we bisect a segment we find its _ _ _ _ _ _ _ _ _ _ _ _ .
    (a) angle   (c) double
    (b) midpoint   (d) vertex

11. If we _ _ _ _ _ _ _ _ _ _ _ _ _ _ a given segment, the new
    segment is three times as long as the given segment.
    (a) double   (c) bisect
    (b) triple   (d) lay off

Key to Geometry® workbooks

Book 1:   Lines and Segments

Book 2:   Circles

Book 3:   Constructions

Book 4:   Perpendiculars

Book 5:   Squares and Rectangles

Book 6:   Angles

Book 7:   Perpendiculars and Parallels, Chords and Tangents, Circles

Book 8:   Triangles, Parallel Lines, Similar Polygons

Answers and Notes for Books 1–3

Answers and Notes for Books 4–6

Answers and Notes for Book 7

Answers and Notes for Book 8

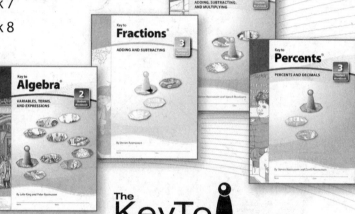

The KeyTo TRACKER™

The Online Companion to the *Key to...*® Workbook Series

Save time, improve learning, and monitor student progress with The Key To Tracker, the online companion to the Key to... workbooks for fractions, decimals, percents, and algebra.

Learn more: www.keypress.com/keyto

Also available in the Key to...® series

Key to Fractions®

Key to Decimals®

Key to Percents®

Key to Algebra®

Key to Metric®

Key to Metric Measurement®

Key Curriculum Press
INNOVATORS IN MATHEMATICS EDUCATION

ISBN 978-0-913684-73-3

90000

9 780913 684733